My Trip
to The

KAZIRANGA
National Park

Sachin Narayan Kadam

Tips for Parents

- Make proper time-table for reading. Allow reading to be a relaxing and enjoyable time.
- Create a comfortable reading space with proper light arrangements to encourage kids for reading.
- Set aside a regular read-aloud with your children. Reading aloud helps your child develop an interest in reading which boosts Imagination Power.
- Initially choose a quality or limited content book that appeals to your child's age and interests.
- Practice read-aloud sessions in a free time.
- Brown Bear publications and Authors' team is new and determined to create quality content for your Reading Rock-Stars.
- Kindly connect with us and we Team B.B. Publications are open for suggestions from parents end. You are welcome if you wish to suggest the topics and subjects for your child's book.
- Have a happy reading session with your kid.
- Let's build together a Reading Rock-Star.

Wooo hooooo ....
Now I am all excited to visit this
Beautiful National Park.

I must clean my desk

And read something more about India

Himalayan Mountains

Deccan Plateau

Desert

Coastal Plains

Tropical Forests

And my destination
the Kaziranga National Park
is situated in
the North Eastern region
of India
in Assam State

Assam is known as the gateway to
north-east India.

It is a land of Blue hills,
Valleys and Rivers

The wife of Lord Curzon (Then Viceroy), Mary Victoria Leiter Curzon established Kaziranga Wildlife Sanctuary in

1905.

Mary Curzon wanted to protect the **decreasing** population of the **Rhinoceros** in the region.

Later In 1950, P.D. Stracey changed the name of **Kaziranga Game Sanctuary** to Kaziranga Wildlife Sanctuary to **avoid Poaching & hunting** activities in the National park.

Kaziranga conservation plays an important role as it is a **home** to two-thirds of the world's great one-horned Rhinoceros population.

It also provides home to some of the world's most endangered species of **Birds** and **wildlife**.

In the year 1985 the Kaziranga National Park was entitled as one of the **UNESCO** **World Heritage Sites in India.**

UNESCO

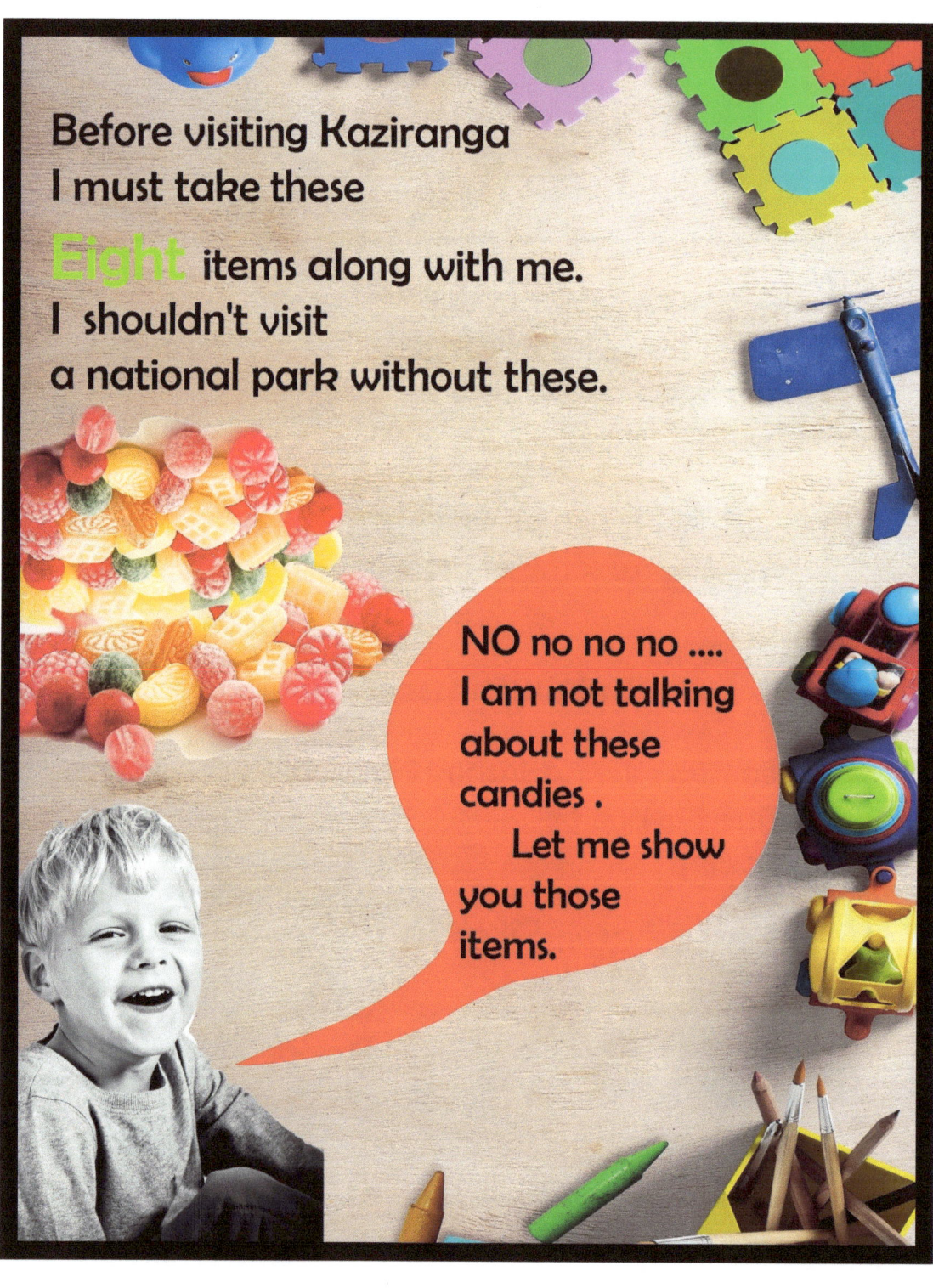

Before visiting Kaziranga
I must take these
Eight items along with me.
I shouldn't visit
a national park without these.

NO no no no ....
I am not talking
about these
candies .
Let me show
you those
items.

The Second important item is

Binocular

- To get a closer look at wonderful and **Colourful** Wildlife.

• Bear Spray

To Keep the Sloth
Bears away

How to **Use a** Bear Spray ?

- • Bear spray has to hit the eyes and
  nose of the bear.

    - • Don't spray when the bear is too far.

- • Aim slightly above the bear's head so
  Gravity will effect the placement of
  the spray.

• Safari Clothes and Hiking Boots

For wildlife safari, wear clothes designed in natural colors which include Pale Green, Khaki, Grey and Brown.

It is quite useful to have light & comfortable shoes. It provides support to your ankles

1. Know your size. ...
2. Try on boots at the end of the day
3. Wear appropriate socks. ...
4. Spend some time in the boots.

Sunscreen
& Sunglasses

• a Paper Map

Print the map of the entire area
beforehand so that
you are not stranded
in case there is no signal or GPS.

• Digital Camera Lenses, Memory cards

You will experience some of the beautiful and amazing sights.

So you need to be prepared to capture it in your camera.

- **Basic First aid kit**

There is a chance you might fall sick or get hurt.

So better to have some remedies at hand - for fever, cold, constipation, a pain reliever gel, some bandages, crepe bandage, etc.

In short, It's very important to keep up your fitness and health routine while enjoying the magic of the Wild.

Finally I have packed my Backpack and I am ready to go. Oh my God ! I am so excited to visit the **Kaziranga** National Park.

Hello Grandpa !

Let me call my Grandpa.

Me: Hello Grandpa. How do you do ?
Grandpa : I am doing great. How about you my child ?
          Are you all set to fly to India for **Kaziranga.**
Me :   Yes Grandpa. I am all set.
          I am going to miss you there.
Grandpa : Aww my Child.... I will miss you too.
Me :   Grandpa . Would you like to give
          me any last minute instruction.
Grandpa : Yes. I was about to give you
          some important instructions
          and few safety tips.

Instructions & Safety Tips

- While on safari drink and carry enough water with you.

- Eat fresh fruits and lighter meal.
Do not pluck and eat wild berries as it can be poisonous sometimes.

- Do not pluck wild flowers. Stay away from wild plants as they may cause skin irritation, itches or Skin allergies.

Instructions & Safety Tips

- Do not go close to wild animals. They all look cute and funny but they are unpredictable and may attack you.

- Do not go too far from your Safari group. There are chances that you can get lost in the forest.

Instructions & Safety Tips

- In case of trouble,
  blow whistle
  if possible

And signal for help
by waving hands at the group members.

If they don't catch your signal
carefully follow fresh foot
prints of the group members if you can.

Instructions & Safety Tips

• Respect the **Nature.**
And Nature will show you some of the
Beautiful sights you have
never seen before.

Wildlife inside Kaziranga :

The Kaziranga National Park houses a huge breed of mammals. The wildlife here is popular because of some rare species of mammals resides here.

One Horned Rhinoceros -

This awesome specie resides with utmost happiness in the Kaziranga National Park.

The fun fact about this mammal is that to battle the heat these rhinos eat during the chilled phases of the day and rest in rivers during the hotter phases.

Wild Buffalo

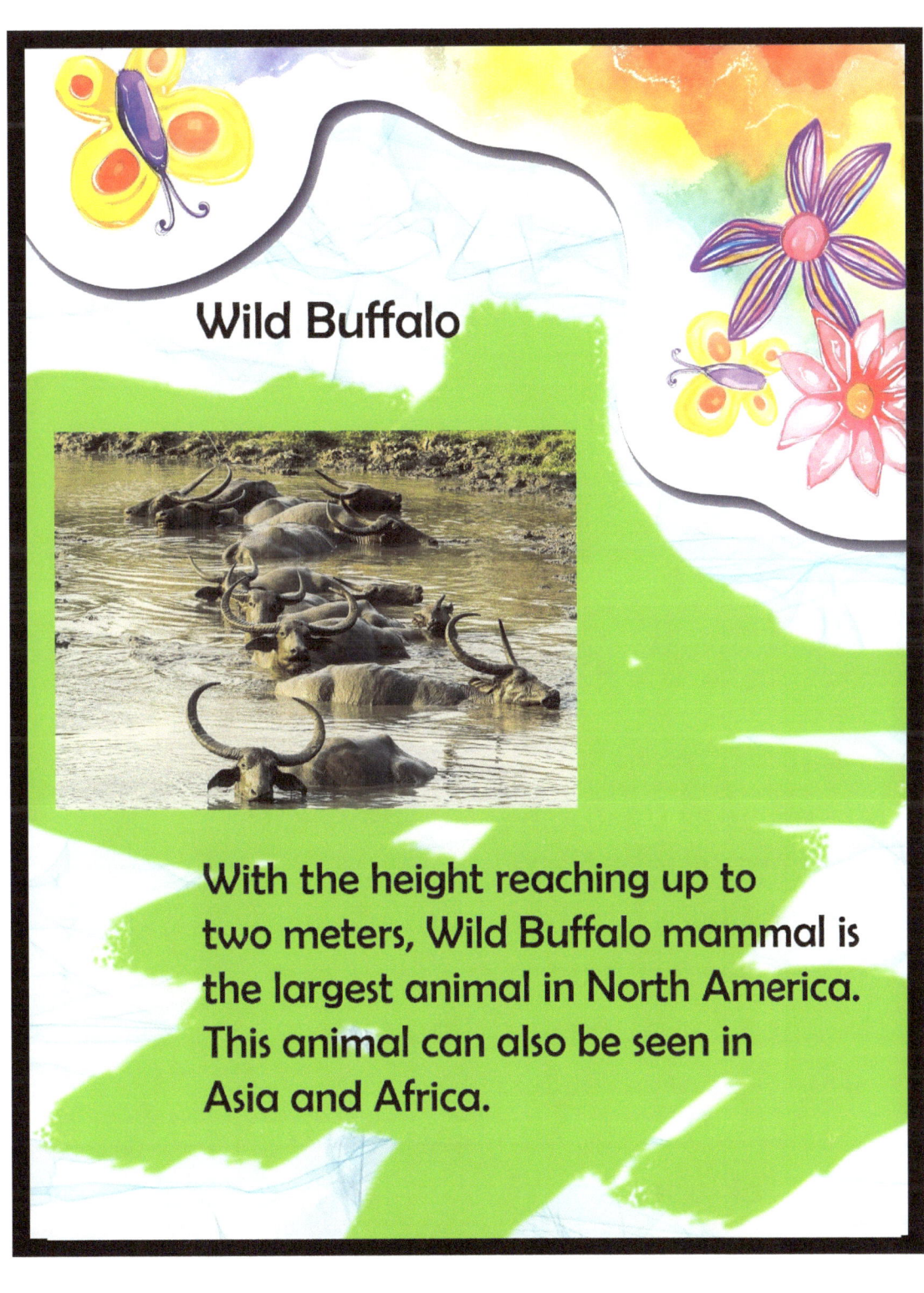

With the height reaching up to two meters, Wild Buffalo mammal is the largest animal in North America. This animal can also be seen in Asia and Africa.

Indian Elephant

The Indian elephant is one of the largest land mammals on the Earth. They have been very important to Asian culture for thousands of years. They have been domesticated and are used for transportation and to move heavy objects.

It feels like you've gone back to the times of the royal when you ride an Elephant

Royal Bengal Tiger

Kaziranga has the highest number of tigers among protected areas in the world.

In the year 2006 Kaziranga was declared as a Tiger Reserve.

The tiger is a unique animal which plays an important role in the health and diversity of an ecosystem. It is a top predator in the food chain and keeps the population of wild ungulates in check, thereby maintaining the balance between prey herbivores and the vegetation upon which they feed.

Eastern Swamp Deer

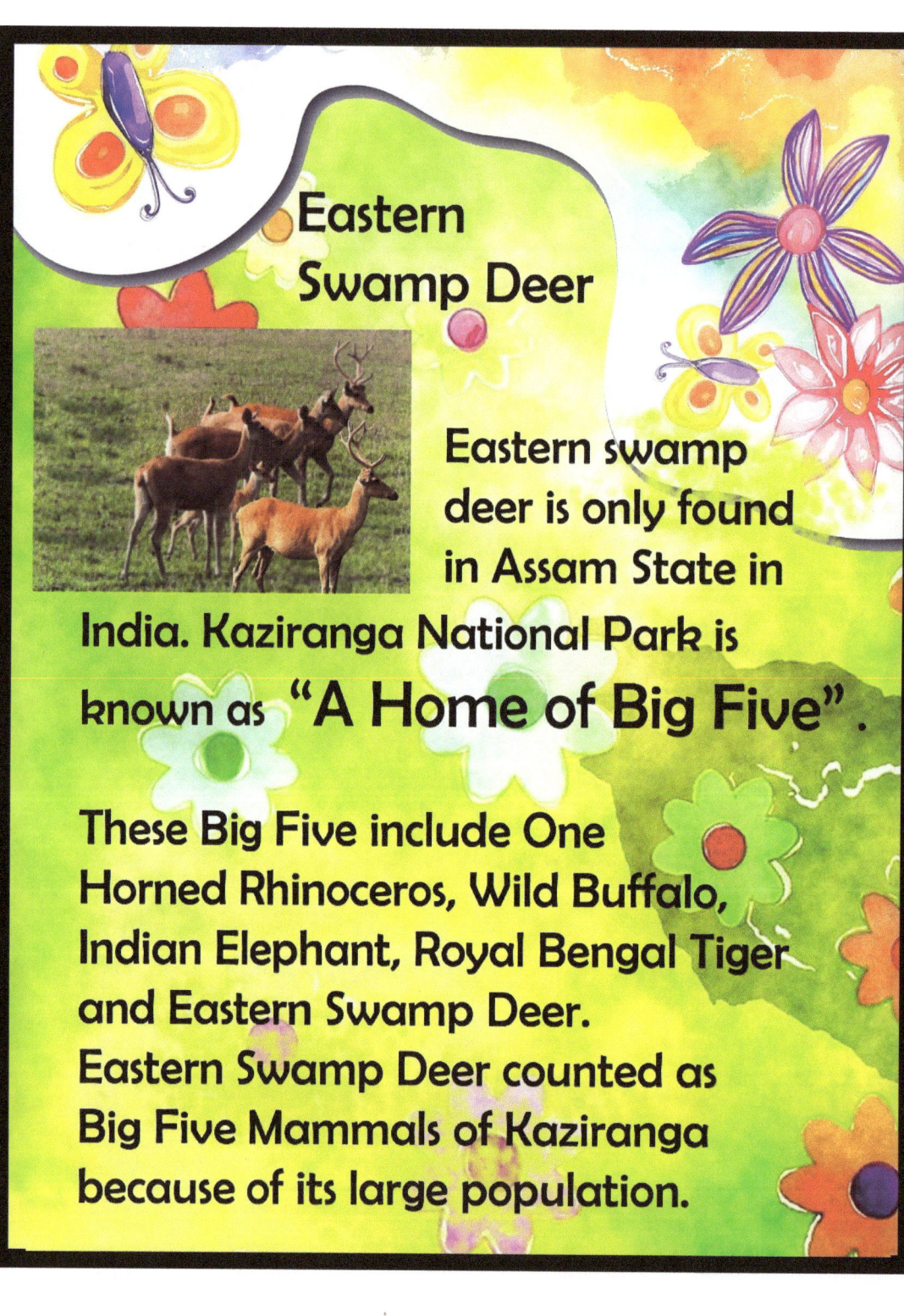

Eastern swamp deer is only found in Assam State in India. Kaziranga National Park is known as **"A Home of Big Five"**.

These Big Five include One Horned Rhinoceros, Wild Buffalo, Indian Elephant, Royal Bengal Tiger and Eastern Swamp Deer. Eastern Swamp Deer counted as Big Five Mammals of Kaziranga because of its large population.

Eastern Mole

This cute mammal originally belongs to America, but can be found here in Kaziranga too.

Hog badger

Another name appears on the vulnerable species of Red list, It is a rarely seen animal, and one can enjoy the view of these species here in Kaziranga.

Ganges and Indus River Dolphin

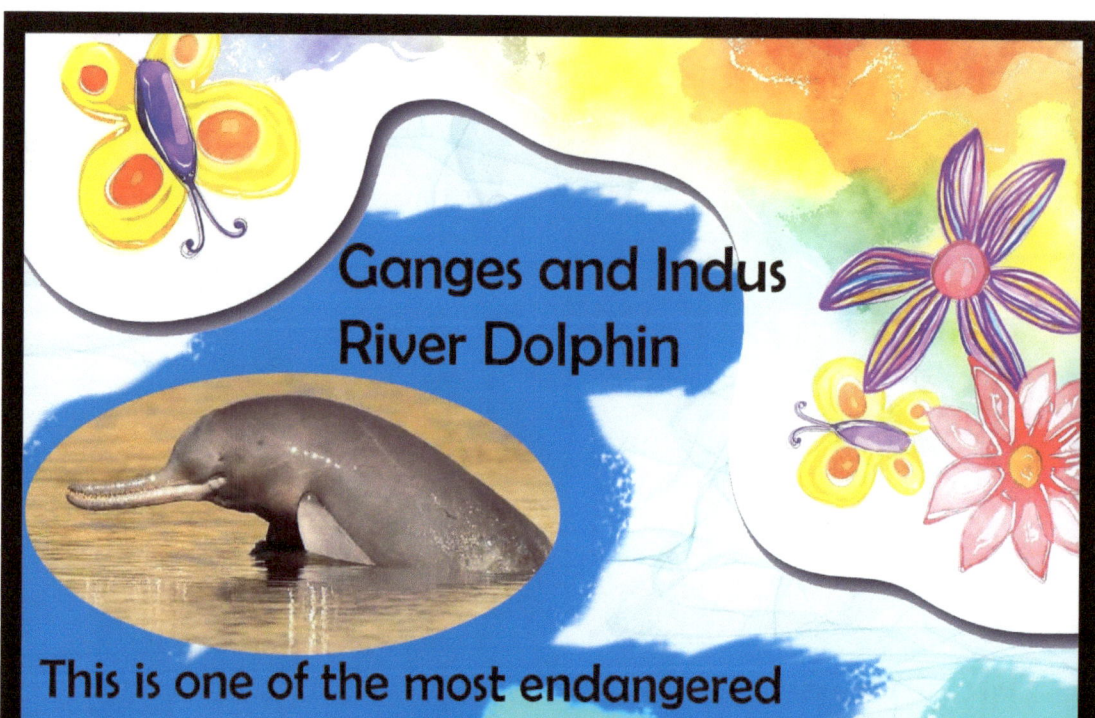

This is one of the most endangered water species found in here.

Large Indian civet and Small Indian civet

Their name has been listed on IUCN Red List.

Indian Pangolin

This poor specie is the **most trafficked** animal and is therefore conserved by the Kaziranga National Park.

Rhesus Macaque

Another soon-going-to extinct species of Old World Monkeys which is conserved by Kaziranga.

Some animals that are endangered
Worldwide are conserved here.
They happily reside here in Peace.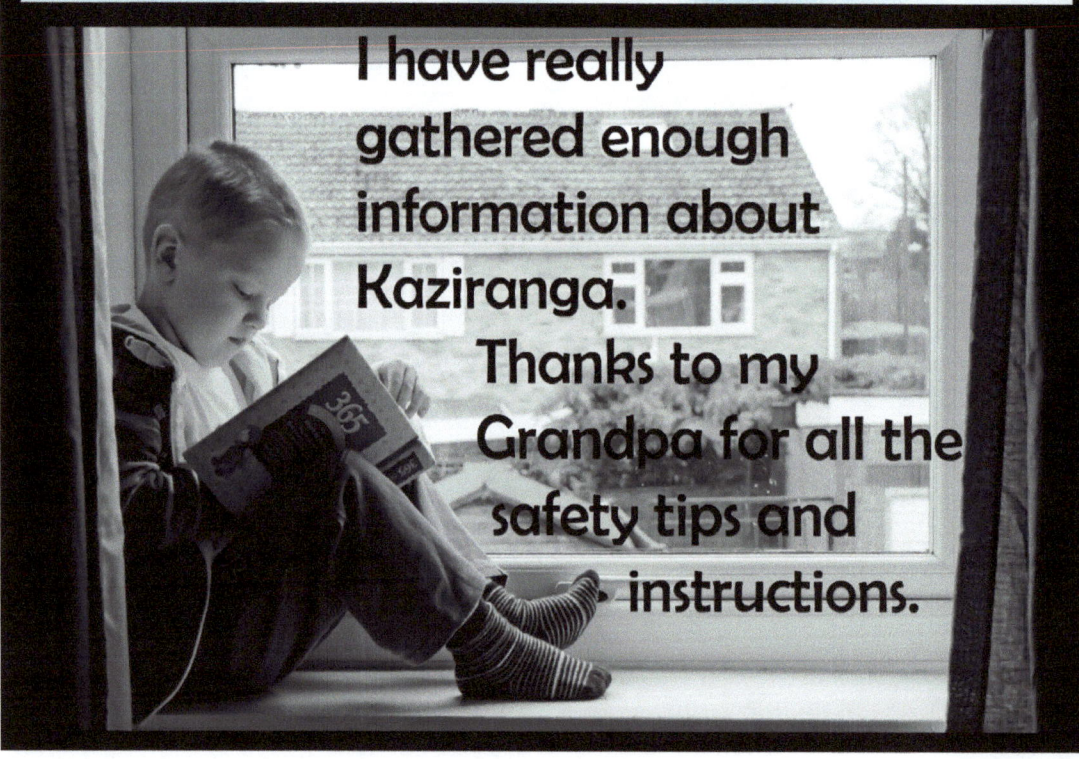

The variety I can see here is amazing,
and therefore this place is
worth a visit.

Enough of reading now.

I must leave for **India**

I have really
gathered enough
information about
Kaziranga.
Thanks to my
Grandpa for all the
safety tips and
instructions.

It's so quiet here. Mom we must take rest now. This area of accomodation is called, 'Nature hunt Eco Camps'.

Let's take rest Mom...

Tomorrow morning we must feel fresh to enjoy the Safari in Park.

Mom.... Will you please just come out. Look at this ..... This is going to be our Safari Ride...... It's Such a Beautiful Elephant.

I am finnally here and I am already loving this place.

My eyes are trying to capture everything through this Camera.

Mom I had a talk with Safari Management Staff and they told me that Kaziranga National Park is divided into the four tourism zones.

1. Central Zone or Kaziranga Zone

2. Western Zone or Bagori Zone

3. Eastern Zone or Agaratoli Zone

4. The Burapahar Zone.

With these many zones spread over nearly 378 square kilometres, now I know why Kaziranga is called as the "Hotspot of Biodiversity".

I am noticing the rich Green Area of the Kaziranga National Park. It consists of the tall grasses like Elephant Grass, Sugarcanes and Spear Grass.

Here are some pictures of types of grass one can find here in this park. Grass feild here is like heaven for the animals who reside here.

An Elephant in Elephant Grass

Sugar Cane

Spear Grass

Water lilies, lotus and water hyacinth enriches the beauty of the Kaziranga National Park.
Water Bodies in the forest are completely covered with the aquatic flora.

< - Water Lilies

Water Hycinth - >

< - Lotus

Thank God I read all the information before coming here at the Kaziranga Park. Thanks to my Grandpa for his valuable safety tips.

Rangers here in the park are so nice. They told me that every year, during the Monsoon season the water of the Brahmaputra River floods the park.

The season continues from June to August. During this season Kaziranga Park becomes inaccessible. November to April is the best time to visit the Park.

So I request everyone to come and visit the Kaziranga National park, make new friends, meet some of the beautiful species of this region, enjoy the beauty of the tropical forests, ride an Elephant and feel like Royal.

Thank you Soo Much Mom for this
Wonderful trip to the Kaziranga.
This is one of the best birthday gifts ever.
I have seen so many beautiful
animals, flowers, Some beautiful
sites of Kaziranga. I am going to store all
the wonderful memories of this trip
forever in my mind .
Thank you Kaziranga !

Author: Sachin Narayan Kadam
Book Cover: u.toptrends
Country: India
City: Mumbai
Email: u.toptrends@gmail.com
               sachinnkadam5@gmail.com

About these Books:

Books published by Sachin Narayan Kadam was designed with specific functions in mind :

Firstly as a reading resource for Preschool, Primary Education and for private Tuitions. They are graded from Simple picture book to the full text . It has a child friendly content. The layout of the book is simple and designed to be fun. The layout and text is designed for easy reading. Parents need to motivate children and inculcate reading habit in their ward. These books serves the purpose.

May we together build a Reading RockStar at your home.